# THIS NOTEBOOK

## BELONGS TO:

*Vintage Fauna*
COLLECTION



Sophie
ALEXANDER
— PUBLISHING —

Made in the USA
Coppell, TX
05 January 2024

27290046R00063